Square Roots

Square Roots

Square Root Expressions
and
Step-by-Step Simplification

Arben Alimi

Square Roots
Square Root Expressions and
Step-by-Step Simplification

Copyright © 2016 Arben Alimi

All rights reserved. Reproduction or translation of any part of this book without the permission and authorization of the copyright owner of this book is unlawful. No part of this publication may be stored, reproduced and transmitted mechanically, electronically, by photocopying or otherwise, except as permitted under United States Copyright laws.

LIMIT OF LIABILTY AND DISCLAIMER OF WARRANTY: THE AUTHOR MAKES NO WARRANTY REGARDING THE ACCURACY AND COMPLETENES OF THE CONTENTS OF THIS PUBLICATION AND THE WARRANTY CANNOT BE CREATED AND EXTENDED BY SALES AND PROMOTIONAL METERIAL. THE AUTHOR SHALL NOT BE LIABLE FOR DAMAGES ARISING HEREFROM.

ISBN-10: 1537597493
ISBN-13: 978-1537597492
CreateSpace Independent Publishing Platform,
North Charleston, SC

Acknowledgments

It is with ineffable gratitude that I acknowledge the immense support and help of my family in writing this book. I owe my deepest gratitude to my caring and loving wife Sevim for managing household and other activities while I was trying to write, calculate and complete this book. I am indebted to my eight year old son Artrim for showing understanding that daddy is busy and does not have enough time to play with him. Finally, the completion of this book would not have been accomplished without the tremendous academic support of my daughter Trina. Her insightful suggestions, meticulous editing and proofreading made this book possible. I share the credit of my work with Trina.

Contents

Introduction to Square Roots 1
Example 1 8
Example 2 11
Example 3 12
Example 4 14
Example 5 15
Example 6 17
Example 7 19
Example 8 19
Example 9 20
Example 10 21
Example 11 22
Example 12 23
Example 13 24
Example 14 25
Example 15 26
Example 16 27
Example 17 27
Example 18 28
Example 19 29
Example 20 29
Example 21 30
Example 22 31
Example 23 32
Example 24 36
Example 25 37
Example 26 38

Example 27 39
Example 28 41
Example 29 42
Example 30 43
Example 31 44
Example 32 45
Example 33 46
Example 34 48
Example 35 49
Example 36 50
Example 37 51
Example 38 52
Example 39 52
Example 40 53
Example 41 54
Example 42 54
Example 43 55
Example 44 56
Example 45 57
Example 46 58
Example 47 59
Example 48 60
Example 49 61
Example 50 62

Preface

You should consider this book as a Personal Tutor who is sitting next to you and is trying hard to make it easy and fun while you are honing your skills on Square Roots. This "Personal Tutor" will not only help you understand how to simplify Square Root expressions, but will increase your confidence to manipulate algebraic expressions in general. As a good Tutor would do, this book does not simply give you the solutions, but shows in detail every step as the examples are being solved. This book-tutor is an affordable alternative to the Real Tutors. This book will be with you all the time and will go over an example as many times as you want, any time you want.

This book is a fruit of father-daughter Math collaboration. As Trina was studying and simplifying Square Root expressions, very often she needed a confirmation that she got it right. Some of the Math expressions were trivial, but some were tricky and required some serious thinking.

As a result of our Math conversations and the fruits that came out of it, we decided to share our knowledge with you. We came up with new examples, wrote down what we were thinking as we solved the problems and put them in this book, so that you and every other student that comes across can benefit the same way Trina benefited from these detailed explanations. Her understanding is the testimony that every solution is easy to follow, detailed, and fully explained.

Being a computer programmer myself, I always think in terms of algorithms, where every step is clearly defined, and not a single step skipped, so that non-intelligent machines, such as computers, can "solve problems" and appear to be intelligent. Similarly, every example in this book is solved step-by-step without a step skipped, so that the student can understand every detail while the expression is simplified

We think it is better to go over the examples in order because we feel that the first ones are easier and can help you build a sound thinking foundation and be better prepared for the later examples, which we feel are more complex.

Introduction to Square Roots

As you already know, there are many kinds of Mathematical Operations: Addition, Subtraction, Multiplication, Division etc.

For every operation we have a Math symbol which we use to denote the operation. As you already know, we use: **+** for Addition, **−** for Subtraction, **×** for Multiplication, and **÷** for Division etc.
But what exactly is a Math operation?

In plain English, a Math operation is the action of taking one or more values (operands), do something with them and produce a new value (result).
For example, you pick any two numbers, let's say: **6** and **2**. You can choose any operation and do something with those two numbers. You can add them, subtract them, multiply, divide them etc., and so produce a new value.

For example:

$$6 + 2 = 8$$
$$6 - 2 = 4$$
$$6 \times 2 = 12$$
$$6 \div 2 = 3$$

Note that all operations above take two values (operands) to produce a new value. That is why they are called binary operations and their symbols are called binary operators.

Wait a second, if they are called binary operations because they take two values (operands), then does that mean there are operations that take **1** value (operand) and produce a new value? You bet there are. Think about Negative numbers. Can you recall which operation can turn a Positive number into a Negative one? It is the Negation operation that does that. The Negation operation shares its symbol with Subtraction, which is the " − " symbol. When you Negate a number you just put the " − " symbol in front of it and the new result value is the negative of the positive operand.

So, Negation is called Unary operation because it takes only one operand (the positive number) and turns it into a negative number. Its symbol " $-$ " is a type of unary operator.

Is there any other unary operation? Of course there is. Think about the cases when a number is multiplied by itself. What did you say about that? You said "square", or "power of **2**", and that is an operation too, called "squaring". When you "square" a value (operand) you produce a new value (result) which is the product of multiplying the operand by itself. So, when you "square" the number **3**, you are multiplying it by itself and the result is **9**:

$$3 \times 3 = 9$$

What is the symbol of this operation? Well, instead of writing 3 × 3, we write:

$$3^2$$

Recall that all these operations have their inverse operations. What is a inverse operation? The inverse operation means the opposite operation which un-does the effect of the first initial operation. Addition and Subtraction are inverse operations of each other because they undo each other:

$$6 + 2 = 8$$

$$8 - 2 = 6$$

Similarly, Multiplication and Division are inverse operations of each other because they undo each other:

$$6 \times 2 = 12$$

$$12 \div 2 = 6$$

How about the square operation? Is there an inverse operation which un-does the "squaring" of a number?

If you say:

$$3^2 = 9$$

then, which operation you will use to bring the result of **9** back to **3** ?

If you said "Square Root" than you are correct, and its symbol is $\sqrt{}$, which is called a Radical.

Let's use this operation to undo the effect of "squaring" the number **3** above and so bring the **9** back to **3**:

$$\sqrt{9} = 3$$

That's it. The Square Root of **9** is **3**. This is so because **3** can be multiplied by **itself** and the result will be **9**. As you can see, this operation is a **Unary** operation, and the radical $\sqrt{}$ is a Unary operator because it takes only one operand, which is called a Radicand.

Let's do some more examples, first square a number and then use the inverse operation of square root to undo it:

$$\text{If } 2^2 = 4 \text{, then } \sqrt{4} = 2$$
$$\text{If } 4^2 = 16 \text{, then } \sqrt{16} = 4$$
$$\text{If } 5^2 = 25 \text{, then } \sqrt{25} = 5$$
$$\text{If } 6^2 = 36 \text{, then } \sqrt{36} = 6$$
$$\text{If } 10^2 = 100 \text{, then } \sqrt{100} = 10$$

Remember this: When you see a $\sqrt{}$ symbol and some number or expression inside it, know that it is just a number that looks different but has the same value. For example when you see a $\sqrt{4}$, think of it like a **2** that looks different. However, sometimes it is best to leave it just like that. For example $\sqrt{2}, \sqrt{3}$ etc. One cannot express these numbers differently because the result will be irrational numbers with decimals that never end:

$$\sqrt{2} = 1.414213562\ldots$$

$$\sqrt{3} = 1.732050807\ldots$$

You can round them up to some decimal, but be aware that that is an approximation, and in Math we would like to be precise. That is why it is preferable to leave them alone in their "Square Root" form like $\sqrt{2}, \sqrt{3}$ etc., because they cannot be written without the Radical. These numbers are called Surds.

Since you are leaving $\sqrt{2}, \sqrt{3}$ the way they are, with the Radical symbol, and not writing them in decimal form, then what do you do if you have to use them in an expression like:

$$\sqrt{2} + 3 - \sqrt{3}$$

Well, since they are still numbers, you can use them in expressions like any other number, but you have to know how to handle them. In other words, you have to know how to add radicals, subtract them, multiply them, divide, square, find square root of square roots, factor out etc., and this is what this book is all about: how to handle and simplify the algebraic expressions where you have square roots, cubic roots etc.

Now that you understand what you will learn from this book, it is time to go over the easy to understand examples and in no time learn how to handle the Radicals like a pro.

Example 1

Simplify: $\sqrt{8}$

Solution

In order to simplify $\sqrt{8}$, think ahead and see whether you can modify the Radicand, the number 8, to a more suitable form which will help you to simplify the entire expression.

You can express the number 8 in different ways, but the simplest way is to express it as product of 4 and 2:

$$\sqrt{8} = \sqrt{4 \times 2}$$

If the Radicand is a product of two numbers, then we can split the expression into two radicals:

$$\sqrt{8} = \sqrt{4} \times \sqrt{2}$$

Express the Radicand 4 as 2 squared:

$$= \sqrt{2^2} \times \sqrt{2}$$

Since "Square Root" and "Square" are inverse operations of each other, then they cancel each other out. This means that we can just remove the radical $\sqrt{}$ and remove the exponent ^{2}. The result will be:

$$= 2 \times \sqrt{2}$$

Or simply:

$$2\sqrt{2}$$

In conclusion:

$$\sqrt{8} = 2\sqrt{2}$$

Check your work

It is very important to always check your work whether you got the correct result, especially for more complex expressions. The simplest way to check for square root expressions is to use the calculator if possible and express them in decimal form. So, using the calculator we will find out that the initial expression $\sqrt{8}$ has the same value in decimal form as the expression after we simplified it such as **2√2**. If the values match, than we did the correct modifications.

So, the initial expression is:

$$\sqrt{8} = 2.82842712474619\ldots$$

and, the simplified expression is:

$$2\sqrt{2} = 2 \times 1.414213562373095\ldots$$
$$= 2.82842712474619\ldots$$

Since the values (the approximations) match, then this is a proof that we did the correct modifications during the simplification.

Note that in this book we are not going to check our work for every example, but we highly encourage you to do so.

Example 2

Simplify: $\sqrt{27}$

Solution

In order to simplify $\sqrt{27}$, think ahead and see whether you can modify the Radicand, the number 27, to a more suitable form which will help you to simplify.

One way to express the number 27 would be as a product of 9 and 3:

$$\sqrt{27} = \sqrt{9 \times 3}$$

If the Radicand is a product of two numbers, then we can split the expression into two radicals:

$$\sqrt{27} = \sqrt{9} \times \sqrt{3}$$

Express the radicand 9 as 3 squared:

$$= \sqrt{3^2} \times \sqrt{3}$$

Since "Square Root" and "Square" are inverse operations of each other, then they cancel each other out. This means that we can just remove the radical $\sqrt{}$ and remove the exponent ^{2}. The result will be:

$$= 3 \times \sqrt{3}$$

Or simply:

$$3\sqrt{3}$$

In conclusion:

$$\sqrt{27} = 3\sqrt{3}$$

Example 3

Simplify: $\sqrt{125}$

Solution

In order to simplify $\sqrt{125}$, think ahead and see whether you can modify the Radicand, the number 125, to a more suitable form which will help you to simplify.

You can express the number 125 in different ways, but a helpful way is to express it as 5^3:

$$\sqrt{125} = \sqrt{5^3}$$

Let's split the Radicand like:

$$\sqrt{5^3} = \sqrt{5^2 \times 5}$$

If the Radicand is a product of two numbers, then we can split the expression into two Radicals:

$$\sqrt{5^2 \times 5} = \sqrt{5^2} \times \sqrt{5}$$

Since "Square Root" and "Square" are inverse operations of each other, then they cancel each other out. This means that we can just remove the radical $\sqrt{}$ and remove the exponent ^{2}. The result will be:

$$= 5 \times \sqrt{5}$$

Or simply:

$$5\sqrt{5}$$

In conclusion:

$$\sqrt{125} = 5\sqrt{5}$$

Example 4

Simplify: $\sqrt{64}$

Solution

In order to simplify $\sqrt{64}$, see whether you can modify the Radicand, the number 64 in a such a way which will help you to simplify further. One way to modify the number 64 is:

$$\sqrt{64} = \sqrt{4 \times 4 \times 4} = \sqrt{4^2 \times 4}$$

Let's split the Radicand like:

$$\sqrt{4^2 \times 4} = \sqrt{4^2} \times \sqrt{4}$$

Since "Square Root" and "Square" are inverse operations of each other, then they cancel each other out. This means that we can just remove the radical $\sqrt{}$ and remove the exponent ^{2}. The result will be:

$$4 \times \sqrt{4}$$

Or simply:

$$4\sqrt{4}$$

Wait a second. This can be simplified even further:

$$4\sqrt{4} = 4 \times \sqrt{2^2} = 4 \times 2 = 8$$

Well, we should have recalled the times table at the beginning of the solution to modify the Radicand 64 to:

$$64 = 8 \times 8 = 8^2$$

Then we would have solved like:

$$\sqrt{64} = \sqrt{8^2} = 8$$

Example 5

Simplify: $\sqrt{8} + \sqrt{18}$

Solution

In order to simplify the whole expression you will need to simplify its terms if possible, and then combine them if possible:

$$\sqrt{8} + \sqrt{18} =$$

Let's split the Radicands like:

$$= \sqrt{4 \times 2} + \sqrt{9 \times 2}$$

$$= \sqrt{2^2 \times 2} + \sqrt{3^2 \times 2}$$

Let's split the terms into two radicals each:

$$= \sqrt{2^2} \times \sqrt{2} + \sqrt{3^2} \times \sqrt{2}$$

Since "Square Root" and "Square" are inverse operations of each other, then they cancel each other out. This means that we can just remove the radical $\sqrt{}$ and remove the exponent 2. The result will be:

$$= 2\sqrt{2} + 3\sqrt{2}$$

After we combine the terms:

$$= 5\sqrt{2}$$

Example 6

$$\text{Simplify:} \quad \sqrt{8} \times \sqrt{18}$$

Solution

In order to simplify the whole expression you will need to simplify its terms if possible, and then combine them if possible:

$$\sqrt{8} \times \sqrt{18} =$$

Let's split the Radicands like:

$$= \sqrt{4 \times 2} \times \sqrt{9 \times 2}$$
$$= \sqrt{2^2 \times 2} \times \sqrt{3^2 \times 2}$$

Let's split the terms into two radicals each:

$$= \sqrt{2^2} \times \sqrt{2} \times \sqrt{3^2} \times \sqrt{2}$$

Since "Square Root" and "Square" are inverse operations of each other, then they cancel each other out. This means that we can just remove the radical $\sqrt{}$ and remove the exponent ^{2}. The result will be:

$$= 2 \times \sqrt{2} \times 3 \times \sqrt{2}$$

After we combine the terms:
$$= 2 \times 3 \times \left(\sqrt{2}\right)^2$$
Cancel out the radicand $\sqrt{}$ and the exponent ^{2}:
$$= 2 \times 3 \times 2 = 12$$

Question: Was there a faster way to simplify the initial expression?
Yes there was:
$$\sqrt{8} \times \sqrt{18} =$$
$$= \sqrt{8 \times 18}$$
$$= \sqrt{144}$$
$$= \sqrt{12^2}$$
$$= 12$$

Example 7

$$\text{Simplify:} \quad \sqrt{2} + \sqrt{3}$$

Solution

Unfortunately there is nothing you can do to simplify the expression. It is in its simplest form.

Example 8

$$\text{Simplify:} \quad \sqrt{2} \times \sqrt{3}$$

Solution

Unfortunately there is nothing you can do to modify the Radicals 2 and 3. All you can do in this expression is to multiply the Radicals:

$$\sqrt{2} \times \sqrt{3} =$$
$$= \sqrt{2 \times 3}$$
$$= \sqrt{6}$$

Example 9

Simplify: $(\sqrt{2})^6$

Solution

Think about getting rid of the radical $\sqrt{}$ and the exponent 6. In order to do that, see whether you can modify the exponent 6. One way would be:

$$(\sqrt{2})^6 =$$
$$= (\sqrt{2})^{2 \times 3}$$
$$= ((\sqrt{2})^2)^3$$

Cancel the radical and the inner exponent:

$$= ((\sqrt{2})^2)^3$$
$$= (2)^3$$
$$= 2 \times 2 \times 2$$
$$= 8$$

Example 10

Simplify:
$$\frac{\sqrt{8}}{\sqrt{2}}$$

Solution

One way to simplify this is to combine the numerator and the denominator under one Radical:

$$\frac{\sqrt{8}}{\sqrt{2}} = \sqrt{\frac{8}{2}} = \sqrt{4} = 2$$

Another way to simplify the expression would be:

$$\frac{\sqrt{8}}{\sqrt{2}} =$$
$$= \frac{\sqrt{4 \times 2}}{\sqrt{2}}$$
$$= \frac{\sqrt{4} \times \sqrt{2}}{\sqrt{2}}$$
$$= \frac{2 \times \sqrt{2}}{\sqrt{2}}$$
$$= 2$$

Example 11

Simplify:

$$\frac{\sqrt{\frac{1}{2}}}{\sqrt{\frac{1}{18}}}$$

Solution

Let's combine the numerator and the denominator under one Radical:

$$\frac{\sqrt{\frac{1}{2}}}{\sqrt{\frac{1}{18}}} = \sqrt{\frac{\frac{1}{2}}{\frac{1}{18}}}$$

$$= \sqrt{\frac{18}{2}}$$

$$= \sqrt{9}$$

$$= 3$$

Example 12

Simplify:

$$\frac{\sqrt{\frac{1}{2}}}{\sqrt{\frac{1}{2}}}$$

Solution

Any number divided by itself equals 1:

$$\frac{\sqrt{\frac{1}{2}}}{\sqrt{\frac{1}{2}}} = 1$$

Even though this example seems to be a simple one, it is important to remember this case and recall it when you solve more complex expressions. Always look for similar terms which cancel each other out (the numbers are being divided by itself) and the whole expression becomes simpler.

Example 13

Simplify:

$$\frac{\sqrt{\frac{1}{4}}}{\sqrt{\frac{1}{18}}}$$

Solution

In general, the rule is: $\frac{\sqrt{a}}{\sqrt{b}} = \sqrt{\frac{a}{b}}$

Let's combine the numerator and the denominator under one Radical:

$$\frac{\sqrt{\frac{1}{4}}}{\sqrt{\frac{1}{18}}} = \sqrt{\frac{\frac{1}{4}}{\frac{1}{18}}} = \sqrt{\frac{18}{4}} =$$

$$= \sqrt{\frac{2 \times 9}{2 \times 2}} = \sqrt{\frac{9}{2}} = \frac{\sqrt{9}}{\sqrt{2}} =$$

$$= \frac{\sqrt{3^2}}{\sqrt{2}} = \frac{3}{\sqrt{2}} =$$

We multiply the numerator and the denominator by $\sqrt{2}$:

$$= \frac{3}{\sqrt{2}} \times \frac{\sqrt{2}}{\sqrt{2}} = \frac{3\sqrt{2}}{(\sqrt{2})^2} = \frac{3\sqrt{2}}{2}$$

Example 14

Simplify:

$$\sqrt{\frac{\sqrt{12}}{\sqrt{3}}}$$

Solution

Simplify first the square root of 12:

$$\sqrt{\frac{\sqrt{12}}{\sqrt{3}}} = \sqrt{\frac{\sqrt{4 \times 3}}{\sqrt{3}}} = \sqrt{\frac{\sqrt{4} \times \sqrt{3}}{\sqrt{3}}}$$

Cancel out $\sqrt{3}$:

$$= \sqrt{\sqrt{4}} = \sqrt{2}$$

Example 15

Simplify:

$$\sqrt{\frac{\sqrt{24}}{\sqrt{2}}}$$

Solution

Let's simplify first the square root of 24:

$$\sqrt{\frac{\sqrt{24}}{\sqrt{2}}} = \sqrt{\frac{\sqrt{2 \times 12}}{\sqrt{2}}} = \sqrt{\frac{\sqrt{2} \times \sqrt{12}}{\sqrt{2}}} =$$

Cancel out $\sqrt{2}$:

$$= \sqrt{\sqrt{12}} = \sqrt{\sqrt{4 \times 3}} = \sqrt{\sqrt{4} \times \sqrt{3}} = \sqrt{2\sqrt{3}}$$

Another way to simplify:

$$\sqrt{\frac{\sqrt{24}}{\sqrt{2}}} = \sqrt{\sqrt{\frac{24}{2}}} = \sqrt{\sqrt{12}} = \sqrt{\sqrt{4 \times 3}} = \sqrt{2\sqrt{3}}$$

Example 16

Simplify:
$$\frac{\sqrt{6} \times \sqrt{6}}{2}$$

Solution

Let's simplify first the numerator:

$$\frac{\sqrt{6} \times \sqrt{6}}{2} = \frac{(\sqrt{6})^2}{2} = \frac{6}{2} = 3$$

Example 17

Simplify:
$$\frac{\sqrt{\frac{4}{5}} \times \sqrt{5}}{2}$$

Solution

Let's simplify first the numerator:

$$\frac{\sqrt{\frac{4}{5}} \times \sqrt{5}}{2} = \frac{\sqrt{\frac{4}{5} \times 5}}{2} = \frac{\sqrt{4}}{2} = \frac{2}{2} = 1$$

Example 18

Simplify:
$$\frac{\sqrt{7} \times \sqrt{6}}{\sqrt{2}}$$

Solution

Let's simplify first the numerator:

$$\frac{\sqrt{7} \times \sqrt{6}}{\sqrt{2}} = \frac{\sqrt{7 \times 6}}{\sqrt{2}} =$$

$$= \frac{\sqrt{7 \times 3 \times 2}}{\sqrt{2}} = \frac{\sqrt{7 \times 3} \times \sqrt{2}}{\sqrt{2}} =$$

Cancel out $\sqrt{2}$:

$$= \sqrt{7 \times 3} = \sqrt{21}$$

Example 19

Simplify:
$$\frac{\sqrt{3}+\sqrt{3}+\sqrt{3}}{3}$$

Solution

Let's simplify first the numerator:

$$\frac{\sqrt{3}+\sqrt{3}+\sqrt{3}}{3}=\frac{3\sqrt{3}}{3}=\sqrt{3}$$

Example 20

Simplify:
$$\frac{\sqrt{3}+\sqrt{3}+4\sqrt{3}}{2\sqrt{3}}$$

Solution

Let's simplify first the numerator:

$$\frac{\sqrt{3}+\sqrt{3}+4\sqrt{3}}{2\sqrt{3}}=\frac{6\sqrt{3}}{2\sqrt{3}}=\frac{6}{2}=3$$

Example 21

Simplify:
$$\frac{\sqrt{8}+\sqrt{18}}{2\sqrt{2}}$$

Solution

Let's simplify first the numerator:

$$\frac{\sqrt{8}+\sqrt{18}}{2\sqrt{2}} = \frac{\sqrt{4\times 2}+\sqrt{9\times 2}}{2\sqrt{2}}$$

$$= \frac{\sqrt{4}\times\sqrt{2}+\sqrt{9}\times\sqrt{2}}{2\sqrt{2}}$$

$$= \frac{2\sqrt{2}+3\sqrt{2}}{2\sqrt{2}}$$

$$= \frac{5\sqrt{2}}{2\sqrt{2}}$$

$$= \frac{5}{2}$$

$$= 2.5$$

Example 22

Simplify:
$$\frac{\sqrt{128}}{\sqrt{32}}$$

Solution

Let's combine the numerator and denominator:

$$\frac{\sqrt{128}}{\sqrt{32}} = \sqrt{\frac{128}{32}} = \sqrt{4} = 2$$

Example 23

Simplify:
$$\frac{2}{\sqrt{12} + \sqrt{48}}$$

Solution

You should already know that if you multiply the numerator and denominator of a fraction with the same number, the fraction does not change its value. It just becomes an equivalent fraction.

In order to simplify this:
$$\frac{2}{\sqrt{12} + \sqrt{48}}$$

we need to find a number or expression, and multiply both, the numerator and denominator, by the same number or expression. This expression and this method should help us to simplify the original expression.

In a two-term-denominator (two-term-polynom), such an expression can be the denominator but with opposite sign of the second term of the denominator. In Math language, this is called a conjugate.

In general, if you have a denominator $a - b$ then you multiply both, the numerator and denominator, by $a + b$, and vice versa, if you have in a denominator $a + b$ than you multiply by $a - b$.

In our case where we have $(\sqrt{12} + \sqrt{48})$ in the denominator, here is how we do it:

$$\frac{2}{\sqrt{12} + \sqrt{48}} \times \frac{\sqrt{12} - \sqrt{48}}{\sqrt{12} - \sqrt{48}}$$

Again, why do you think you can multiply by that expression? Think about this: "If you multiply our initial expression by 1, our expression will not change, right? That's exactly what we did:

$$\frac{\sqrt{12} - \sqrt{48}}{\sqrt{12} - \sqrt{48}} = 1$$

If you still ask how is the above expression equal to 1, well, that is because anything divided by itself equals 1.

So, we did not change the value of our expression, but we changed the way it looks. In other words we created an equivalent expression so that we can simplify our initial expression.

Now, why will this help us to simplify? Recall the algebraic formula for $a^2 - b^2$:

$$(a - b) \times (a + b) =$$
$$= a \times a + a \times b - b \times a - b \times b =$$
$$= a^2 - b^2$$

Let's simplify now:

$$\frac{2}{\sqrt{12} + \sqrt{48}} =$$

$$= \frac{2}{\sqrt{12} + \sqrt{48}} \times \frac{\sqrt{12} - \sqrt{48}}{\sqrt{12} - \sqrt{48}}$$

$$= \frac{2 \times (\sqrt{12} - \sqrt{48})}{(\sqrt{12} + \sqrt{48}) \times (\sqrt{12} - \sqrt{48})}$$

We will now modify the denominator according to the formula for $a^2 - b^2$, which we explained above. The expression will look now like:

$$= \frac{2 \times (\sqrt{12} - \sqrt{48})}{(\sqrt{12})^2 - (\sqrt{48})^2} =$$

Cancel the Radical and Exponent in the denominator, and multiply the terms in the numerator:

$$= \frac{2\sqrt{12} - 2\sqrt{48}}{12 - 48} =$$

Simplify the terms in the numerator:

$$= \frac{2\sqrt{4 \times 3} - 2\sqrt{16 \times 3}}{-36}$$

$$= \frac{2 \times \sqrt{4} \times \sqrt{3} - 2 \times \sqrt{16} \times \sqrt{3}}{-36}$$

$$= \frac{2 \times 2 \times \sqrt{3} - 2 \times 4 \times \sqrt{3}}{-36}$$

$$= \frac{4\sqrt{3} - 8\sqrt{3}}{-36}$$

$$= \frac{-4\sqrt{3}}{-36}$$

$$= \frac{-4}{-36}\sqrt{3} = \frac{1}{9}\sqrt{3}$$

Example 24

Simplify:
$$\frac{4}{\sqrt{8}}$$

Solution

Let's multiply the numerator and denominator of the fraction with the same number which will help us get rid of the Radical in the denominator. Can you guess which number would do the job?

It is $\sqrt{8}$. Let's see how:

$$\frac{4}{\sqrt{8}} = \frac{4}{\sqrt{8}} \times \frac{\sqrt{8}}{\sqrt{8}} = \frac{4 \times \sqrt{8}}{\left(\sqrt{8}\right)^2} =$$

Cancel out $\sqrt{}$ and ^{2} in the denominator, and simplify $\sqrt{8}$ in the numerator:

$$= \frac{4 \times \sqrt{4 \times 2}}{8} = \frac{4 \times \sqrt{4} \times \sqrt{2}}{8} =$$

$$= \frac{4 \times 2 \times \sqrt{2}}{8} = \frac{8 \times \sqrt{2}}{8} = \sqrt{2}$$

Example 25

Simplify:
$$\frac{4\sqrt{3}}{\sqrt{12}}$$

Solution

If you take a closer look, you would recognize that the Radicand 3 in the numerator is a factor of the Radicand 12 in the denominator. This should be a hint that there might be possible to factor something out and then cancel something so that the expression gets simplified a bit.

Let's modify the denominator:

$$\frac{4\sqrt{3}}{\sqrt{12}} = \frac{4\sqrt{3}}{\sqrt{4 \times 3}} = \frac{4\sqrt{3}}{\sqrt{4} \times \sqrt{3}} =$$

$$= \frac{4\sqrt{3}}{2\sqrt{3}} = 2$$

We Canceled out $\sqrt{3}$ and divided 4 by 2.

Example 26

Simplify:
$$\frac{6}{2-\sqrt{2}}$$

Solution

Let's multiply the numerator and denominator of the fraction with the same number which will help us get rid of the Radical in the denominator.

$$\frac{6}{2-\sqrt{2}} = \frac{6}{2-\sqrt{2}} \times \frac{2+\sqrt{2}}{2+\sqrt{2}} = \frac{6 \times (2+\sqrt{2})}{(2-\sqrt{2}) \times (2+\sqrt{2})} =$$

Recall that: $(a-b)(a+b) = a^2 - b^2$. Let's use this for the expression in the denominator:

$$= \frac{6 \times (2+\sqrt{2})}{2^2 - (\sqrt{2})^2} =$$

$$= \frac{12 + 6\sqrt{2}}{4 - 2} = \frac{12 + 6\sqrt{2}}{2} =$$

$$= \frac{12}{2} + \frac{6\sqrt{2}}{2} = 6 + 3\sqrt{2}$$

Example 27

Simplify:
$$\frac{1 - \sqrt{2}}{1 + \sqrt{2}}$$

Solution

Let's multiply the numerator and denominator of the fraction with the same number which will help us get rid of the Radical in the denominator.

$$\frac{1 - \sqrt{2}}{1 + \sqrt{2}} = \frac{1 - \sqrt{2}}{1 + \sqrt{2}} \times \frac{1 - \sqrt{2}}{1 - \sqrt{2}} =$$

$$= \frac{(1-\sqrt{2}) \times (1-\sqrt{2})}{(1+\sqrt{2}) \times (1-\sqrt{2})} =$$

Recall that: $(a-b)(a+b) = a^2 - b^2$. Let's use this for the expression in the denominator:

$$= \frac{(1-\sqrt{2}) \times (1-\sqrt{2})}{1^2 - (\sqrt{2})^2} =$$

Let's multiply the terms in the numerator:

$$= \frac{1 - \sqrt{2} - \sqrt{2} + (\sqrt{2})^2}{1 - 2} =$$

Let's combine the terms in the numerator:

$$= \frac{1 - 2\sqrt{2} + 2}{-1} = \frac{3 - 2\sqrt{2}}{-1} =$$

$$= \frac{3}{-1} - \frac{2\sqrt{2}}{-1}$$

$$= -3 + 2\sqrt{2}$$

Example 28

Simplify:
$$\frac{-1-\sqrt{2}}{1+\sqrt{2}}$$

Solution

Factor out -1 from the numerator:

$$\frac{-1-\sqrt{2}}{1+\sqrt{2}} = \frac{-1\times(1+\sqrt{2})}{1+\sqrt{2}} =$$

Not necessary, but let's remove -1 out of the fraction for better visibility:

$$= -1 \times \frac{1+\sqrt{2}}{1+\sqrt{2}} =$$

Any number divided by itself equals 1, so does the number $(1+\sqrt{2})$ too:

$$= -1 \times 1 = -1$$

Example 29

Simplify:
$$\frac{9}{\sqrt{5} - \sqrt{2}}$$

Solution

Let's multiply the numerator and denominator of the fraction with the same number which will help us get rid of the Radical in the denominator.

$$\frac{9}{\sqrt{5} - \sqrt{2}} = \frac{9}{\sqrt{5} - \sqrt{2}} \times \frac{\sqrt{5} + \sqrt{2}}{\sqrt{5} + \sqrt{2}} =$$

$$= \frac{9 \times (\sqrt{5} + \sqrt{2})}{(\sqrt{5} - \sqrt{2}) \times (\sqrt{5} + \sqrt{2})} =$$

Recall that: $(a - b)(a + b) = a^2 - b^2$. Let's use this for the expression in the denominator, and multiply the terms in numerator:

$$= \frac{9\sqrt{5} + 9\sqrt{2}}{\left(\sqrt{5}\right)^2 - \left(\sqrt{2}\right)^2} =$$

$$= \frac{9\sqrt{5} + 9\sqrt{2}}{5 - 2} = \frac{9\sqrt{5} + 9\sqrt{2}}{3} =$$

$$= \frac{9\sqrt{5}}{3} + \frac{9\sqrt{2}}{3} = 3\sqrt{5} + 3\sqrt{2}$$

Example 30

Simplify:
$$\sqrt[3]{27} + \sqrt[4]{16}$$

Solution

Recall that:

$27 = 3 \times 3 \times 3 = 3^3$, and $16 = 2 \times 2 \times 2 \times 2 = 2^4$

Let's use this in our expression:

$$\sqrt[3]{27} + \sqrt[4]{16} =$$
$$= \sqrt[3]{3^3} + \sqrt[4]{2^4} =$$
$$= 3 + 2 = 5$$

Example 31

Simplify:
$$\sqrt[3]{8} + \sqrt{8}$$

Solution

$$\sqrt[3]{8} + \sqrt{8} =$$
$$= \sqrt[3]{2^3} + \sqrt{2^2 \times 2} =$$
$$= 2 + 2\sqrt{2}$$

Example 32

Simplify:
$$\frac{\sqrt[3]{16} + \sqrt{8}}{2}$$

Solution

Let's modify the radicands in the numerator in a such a way so that we can factor something out:

$$\frac{\sqrt[3]{16} + \sqrt{8}}{2} = \frac{\sqrt[3]{8 \times 2} + \sqrt{4 \times 2}}{2}$$

$$= \frac{\sqrt[3]{2^3 \times 2} + \sqrt{2^2 \times 2}}{2}$$

$$= \frac{\sqrt[3]{2^3} \times \sqrt[3]{2} + \sqrt{2^2} \times \sqrt{2}}{2} =$$

Cancel out $\sqrt[3]{}$ and the exponent 3, and $\sqrt{}$ the exponent 2:

$$= \frac{2 \times \sqrt[3]{2} + 2 \times \sqrt{2}}{2}$$

$$= \frac{2 \times (\sqrt[3]{2} + \sqrt{2})}{2}$$

$$= \sqrt[3]{2} + \sqrt{2}$$

Example 33

Simplify:
$$\frac{\sqrt{2} + 2\sqrt[3]{2}}{\sqrt{2}}$$

Solution

Let's multiply the numerator and the denominator with the same number which will help us get rid of the Radical in the denominator. Such a number would be $\sqrt{2}$:

$$\frac{\sqrt{2} + 2\sqrt[3]{2}}{\sqrt{2}} = \frac{\sqrt{2} + 2\sqrt[3]{2}}{\sqrt{2}} \times \frac{\sqrt{2}}{\sqrt{2}}$$

$$= \frac{(\sqrt{2} + 2\sqrt[3]{2}) \times \sqrt{2}}{(\sqrt{2})^2}$$

$$= \frac{(\sqrt{2})^2 + 2\sqrt{2}\sqrt[3]{2}}{2}$$

$$= \frac{2 + 2\sqrt{2}\sqrt[3]{2}}{2} =$$

Factor 2 out in the numerator:

$$= \frac{2 \times (1 + \sqrt{2}\sqrt[3]{2})}{2}$$

$$= 1 + \sqrt{2}\sqrt[3]{2}$$

Remember this: $\sqrt[n]{x^m} = x^{\frac{m}{n}}$, or $\sqrt[n]{x} = x^{\frac{1}{n}}$.

In our case we have:

$$= 1 + 2^{\frac{1}{2}} \times 2^{\frac{1}{3}} =$$

$$= 1 + 2^{\frac{1}{2}+\frac{1}{3}} =$$

$$= 1 + 2^{\frac{5}{6}} =$$

$$= 1 + \sqrt[6]{2^5} =$$

$$= 1 + \sqrt[6]{32}$$

Example 34

Simplify:

$$\frac{\sqrt{8} + \sqrt[3]{16} + \sqrt[4]{32} + \sqrt[5]{64}}{2}$$

Solution

Let's first modify the radicands and factor something out:
$$\frac{\sqrt{8} + \sqrt[3]{16} + \sqrt[4]{32} + \sqrt[5]{64}}{2} =$$

$$= \frac{\sqrt{4 \times 2} + \sqrt[3]{8 \times 2} + \sqrt[4]{16 \times 2} + \sqrt[5]{32 \times 2}}{2} =$$

$$= \frac{\sqrt{2^2 \times 2} + \sqrt[3]{2^3 \times 2} + \sqrt[4]{2^4 \times 2} + \sqrt[5]{2^5 \times 2}}{2} =$$

Let's factor out 2^n where $n = 2,3,4,5$:

$$= \frac{2\sqrt{2} + 2\sqrt[3]{2} + 2\sqrt[4]{2} + 2\sqrt[5]{2}}{2}$$

$$= \frac{2(\sqrt{2} + \sqrt[3]{2} + \sqrt[4]{2} + \sqrt[5]{2})}{2}$$

$$= \sqrt{2} + \sqrt[3]{2} + \sqrt[4]{2} + \sqrt[5]{2}$$

Example 35

Simplify:

$$\sqrt[3]{\sqrt{2}}$$

Solution

We should remember this:

$$\sqrt{x} = x^{\frac{1}{2}}$$

$$\sqrt[3]{x} = x^{\frac{1}{3}}$$

$$\sqrt[n]{x} = x^{\frac{1}{n}}$$

Using this notation, we can combine the roots like:

$$\sqrt[3]{\sqrt{2}} = \sqrt[3]{2^{\frac{1}{2}}} = \left(2^{\frac{1}{2}}\right)^{\frac{1}{3}} =$$

Multiply the exponents, and write the expression with a radical:

$$= 2^{\frac{1}{2} \times \frac{1}{3}} = 2^{\frac{1}{6}} = \sqrt[6]{2}$$

Example 36

Simplify:

$$\sqrt[3]{\sqrt[4]{x}}$$

Solution

We should not forget this:

$$\sqrt[n]{x} = x^{\frac{1}{n}}$$

Using this rule, we can combine the nested Radicals:

$$\sqrt[3]{\sqrt[4]{x}} = \left(x^{\frac{1}{4}}\right)^{\frac{1}{3}} =$$

We multiply the exponents, and we write the expression with a radical:

$$= x^{\frac{1}{4} \times \frac{1}{3}} = x^{\frac{1}{12}} = \sqrt[12]{x}$$

So, the general rule is:

$$\sqrt[a]{\sqrt[b]{x}} = \sqrt[a \times b]{x}$$

Example 37

Simplify:

$$\sqrt[4]{x^2}$$

Solution

Once again, let's not forget this rule:

$$\sqrt[n]{x} = x^{\frac{1}{n}}$$

Using this rule, we can simplify like this:

$$\sqrt[4]{x^2} = (x^2)^{\frac{1}{4}} = x^{2 \times \frac{1}{4}} = x^{\frac{1}{2}} = \sqrt{x}$$

In general:

$$\sqrt[b]{x^a} = x^{\frac{a}{b}}$$

Using this general formula:

$$\sqrt[4]{x^2} = x^{\frac{2}{4}} = x^{\frac{1}{2}} = \sqrt{x}$$

Example 38

Simplify: $\sqrt[4]{4}$

Solution

Express the Radicand 4 in exponential form:

$$\sqrt[4]{4} = \sqrt[4]{2^2} =$$

Again, do not forget this: $\sqrt[n]{x} = x^{\frac{1}{n}}$

$$= (2^2)^{\frac{1}{4}} = 2^{2 \times \frac{1}{4}} = 2^{\frac{1}{2}} = \sqrt{2}$$

Example 39

Simplify:

$$\sqrt[3]{x}\sqrt{x}$$

Solution

Express first the Radicals in exponential form:

$$\sqrt[3]{x}\sqrt{x} = x^{\frac{1}{3}} \times x^{\frac{1}{2}} =$$
$$= x^{\frac{1}{3}+\frac{1}{2}} = x^{\frac{5}{6}} = \sqrt[6]{x^5}$$

Example 40

Simplify:
$$\sqrt[3]{x^6}\sqrt{x^4}$$

Solution

Express first the Radicals in exponential form:

$$\sqrt[3]{x^6}\sqrt{x^4} = x^{\frac{6}{3}} \times x^{\frac{4}{2}} = x^2 \times x^2 = x^{2+2} = x^4$$

Example 41

Simplify:
$$\sqrt[3]{x^6} + \sqrt{x^4}$$

Solution

Express first the Radicals in exponential form:

$$\sqrt[3]{x^6} + \sqrt{x^4} = x^{\frac{6}{3}} + x^{\frac{4}{2}} = x^2 + x^2 = 2x^2$$

Example 42

Simplify:
$$\frac{\sqrt[3]{x^9}}{\sqrt{x^6}}$$

Solution

Express first the Radicals in exponential form:

$$\frac{\sqrt[3]{x^9}}{\sqrt{x^6}} = \frac{x^{\frac{9}{3}}}{x^{\frac{6}{2}}} = \frac{x^3}{x^3} = 1$$

Example 43

Simplify:

$$\frac{\sqrt[4]{x^2}}{\sqrt{x}}$$

Solution

Express first the Radical in the numerator in exponential form:

$$\frac{\sqrt[4]{x^2}}{\sqrt{x}} = \frac{x^{\frac{2}{4}}}{\sqrt{x}} = \frac{x^{\frac{1}{2}}}{\sqrt{x}} = \frac{\sqrt{x}}{\sqrt{x}} = 1$$

Example 44

Simplify:

$$\frac{\sqrt{x}}{\sqrt[3]{x}}$$

Solution

Express first the Radical in the numerator in exponential form:

$$\frac{\sqrt{x}}{\sqrt[3]{x}} = \frac{x^{\frac{1}{2}}}{x^{\frac{1}{3}}} =$$

Let's convert the fraction in multiplication:

$$= x^{\frac{1}{2}} \times \frac{1}{x^{\frac{1}{3}}} =$$

There is a rule about negative exponents: $x^{-n} = \frac{1}{x^n}$

Using this rule we can modify the expression like:

$$= x^{\frac{1}{2}} \times x^{-\frac{1}{3}} = x^{\frac{1}{2}+\left(-\frac{1}{3}\right)} =$$

$$= x^{\frac{1}{2}-\frac{1}{3}} = x^{\frac{1}{6}} = \sqrt[6]{x}$$

Example 45

Simplify:

$$x^{-2}\sqrt{x^4}$$

Solution

Don't forget the rule about negative exponents: $x^{-n} = \frac{1}{x^n}$

Now, using this rule we will modify the first term:

$$x^{-2}\sqrt{x^4} = \frac{1}{x^2}\sqrt{x^4} = \frac{\sqrt{x^4}}{x^2} =$$

$$= \frac{x^{\frac{4}{2}}}{x^2} = \frac{x^2}{x^2} = 1$$

Example 46

Simplify:

$$x^{-\frac{1}{2}}\sqrt{x^3}$$

Solution

Once again, the rule about negative exponents: $x^{-n} = \frac{1}{x^n}$

Now, using this rule we will modify the first term:

$$x^{-\frac{1}{2}}\sqrt{x^3} = \frac{1}{x^{\frac{1}{2}}}\sqrt{x^3} = \frac{\sqrt{x^3}}{x^{\frac{1}{2}}} =$$

$$= \frac{\sqrt{x^3}}{\sqrt{x}} = \frac{\sqrt{x^{2+1}}}{\sqrt{x}} = \frac{\sqrt{x^2 \times x}}{\sqrt{x}} =$$

$$= \frac{\sqrt{x^2} \times \sqrt{x}}{\sqrt{x}} = \frac{x\sqrt{x}}{\sqrt{x}} = x$$

Example 47

Simplify:

$$\frac{\sqrt[3]{\sqrt{x^3}}}{\sqrt{x}}$$

Solution

We should first simplify the numerator:

$$\frac{\sqrt[3]{\sqrt{x^3}}}{\sqrt{x}} = \frac{(\sqrt{x^3})^{\frac{1}{3}}}{\sqrt{x}} = \frac{\left((x^3)^{\frac{1}{2}}\right)^{\frac{1}{3}}}{\sqrt{x}} =$$

We multiply the exponents in the numerator:

$$= \frac{x^{3 \times \frac{1}{2} \times \frac{1}{3}}}{\sqrt{x}} = \frac{x^{\frac{3}{6}}}{\sqrt{x}} = \frac{x^{\frac{1}{2}}}{\sqrt{x}} = \frac{\sqrt{x}}{\sqrt{x}} = 1$$

Example 48

Simplify:

$$\left(\sqrt[4]{\sqrt[3]{\sqrt{x}}}\right)^{24}$$

Solution

We should first express the radicals as exponents:

$$\left(\sqrt[4]{\sqrt[3]{\sqrt{x}}}\right)^{24} = \left(x^{\frac{1}{2} \times \frac{1}{3} \times \frac{1}{4}}\right)^{24}$$

$$= \left(x^{\frac{1}{24}}\right)^{24}$$

$$= x^{\frac{24}{24}}$$

$$= x$$

Example 49

Simplify:

$$\frac{\sqrt{x^2 + 2xy + y^2}}{x + y}$$

Solution

In order to simplify our expression we need to recall the following Algebra rule:

$$(x + y)^2 = (x + y) \times (x + y)$$
$$= x^2 + xy + yx + y^2$$
$$= x^2 + 2xy + y^2$$

Now, using this rule we will modify the Radicand in the numerator:

$$\frac{\sqrt{x^2 + 2xy + y^2}}{x + y} = \frac{\sqrt{(x + y)^2}}{x + y} =$$

Cancel out the radical $\sqrt{}$ and the exponent ^{2}, and recall that any number divided by itself equals 1:

$$= \frac{x + y}{x + y} = 1$$

Example 50

Simplify:

$$\frac{x^2 - y^2}{\sqrt{x^2 - 2xy + y^2}}$$

Solution

In order to simplify our expression we need to think what we can do in the numerator and what we can do in the denominator in the hope to find a way to cancel something out.

Let's start with the denominator. Similar to the previous example, we should recall the following Algebra rule:

$$(x - y)^2 = (x - y) \times (x - y)$$
$$= x^2 - xy - yx + y^2$$
$$= x^2 - 2xy + y^2$$

Now, using this rule we will modify the Radicand in the denominator:

$$\frac{x^2 - y^2}{\sqrt{x^2 - 2xy + y^2}} = \frac{x^2 - y^2}{\sqrt{(x - y)^2}} =$$

Cancel out the radical $\sqrt{}$ and the exponent ^{2}:

$$= \frac{x^2 - y^2}{x - y} =$$

Is there anything we can do to simplify the numerator? Of course there is. Recall the following Algebra rule:

$$(x + y) \times (x - y) = x^2 - xy + yx - y^2$$
$$= x^2 - y^2$$

Now, using this rule we will modify the numerator:

$$= \frac{x^2 - y^2}{x - y} = \frac{(x + y)(x - y)}{x - y} = x + y$$

So, we found out:

$$\frac{x^2 - y^2}{\sqrt{x^2 - 2xy + y^2}} = x + y$$

www.ingramcontent.com/pod-product-compliance
Lightning Source LLC
Chambersburg PA
CBHW060417190526
45169CB00002B/938